ANCIENT

A DOZEN LOST WORLDS BASED ON THE GEOLOGY OF THE BIGHORN BASIN

WYOMING

KIRK JOHNSON AND **WILL CLYDE**

PAINTINGS BY JAN VRIESEN · ILLUSTRATIONS BY MARJORIE LEGGITT

FULCRUM

DENVER MUSEUM OF
NATURE & SCIENCE
DMNS.ORG · MONTVIEW & COLORADO BLVD. · 303.370.6000

Library of Congress Cataloging-in-Publication Data
Johnson, Kirk R.
  Ancient Wyoming : a dozen lost worlds based on the geology of the Bighorn Basin / Kirk Johnson & Will Clyde.
    pages cm
  Audience: Age 13.
  Audience: Grade 7 to 8.
  ISBN 978-1-936218-09-7
1. Geology–Wyoming–Juvenile literature. 2. Geology–Bighorn Basin (Mont. and Wyo.)–Juvenile literature.
3. Paleontology–Wyoming–Juvenile literature. 4. Paleontology–Bighorn Basin (Mont. and Wyo.)–Juvenile literature.
5. Bighorn Basin (Mont. and Wyo.)–Juvenile literature.
I. Clyde, Will (William C.) II. Title.
  QE181.J64 2015
  557.87–dc23

Printed in the United States.

Design by: Craig Wheeless, Rainy Day Designs

Published by:
Fulcrum Publishing
4690 Table Mountain Drive, Suite 100
Golden, Colorado 80403
(800) 992-2908 • (303) 277-1623
www.fulcrumbooks.com

The Denver Museum of Nature & Science
2001 Colorado Boulevard
Denver, Colorado 80205
www.DMNS.org

Image credits: Denver Museum of Nature & Science (background photo, title page and imprint page; upper left inset on p. 24); Ken Rose (inset on back cover and imprint page); Kirk Johnson (p. 4; p. 6; p. 9; background photo, p. 10; inset photos, p. 13; p. 16; p. 17; p. 20; upper right and lower images on p. 24; p. 25; p. 28; p. 29; p. 32; bottom photo on p. 33; inset photo on p. 36; p. 37, p. 40; p. 41; lower left photo on p. 44; lower right photo on p. 45; lower right photo on p. 49; p. 52; inset photo on p. 57; p. 60; p. 61); Will Clyde (p. 7; formation images, p. 10; paleobotanist holding a core barrel on p. 12; p. 48; p. 53; p. 57; map on p. 62); Scott Wing (p. 3; p. 5); Shutterstock (p. 8); Marjorie Leggitt, Jan Vriesen, and Craig Wheeless (figures on pp. 10, 14, 18, 22, 26, 30, 34, 38, 42, 46, 50, 54, 58); Marjorie Leggitt (cross section on p. 11); Ron C. Blakey (globe images on pp. 11, 14, 18, 22, 26, 30, 34, 38, 42, 46, 50, 54, 58); University of Wyoming archives (p. 12); Jan Vriesen (illustrations on pp. 15, 19, 23, 27, 31, 35, 39, 43, 47, 51, 55, 59); Don Hurlbert (top photo on p. 21); Ray Troll (illustration on p. 21); Laura Viette (inset photo on p. 21); David Lovelace (inset photos on p. 33); American Museum of Natural History (p. 36); Scott Wing and Amy Morey (leaf images, pp. 44–45); Eric De Bast (left photo on p. 49); Jon Bloch (upper right inset photo on p. 49); Roy Campbell (p. 56); Joe Liccardi (upper left map on p. 61); Malin Clyde (p. 64)

CONTENTS

Opposite: Skull of *Wyolestes*, an extinct carnivore from the Willwood Formation. Complete fossil skulls are rare, and this one even has part of its neck attached.

Above: Exposures of the Paleocene-Eocene boundary in the Willwood Formation at Polecat Bench near Powell, Wyoming

Preface

A distinctive layer of rock is often called a *formation*, and every formation has three dimensions: thickness, width, and length. They can be as thin as a few dozen feet or as thick as a mile or two, and their width and length can vary from a few miles to hundreds of miles. Imagine them as giant stone pancakes. And like pancakes, you find them in stacks. A canyon is a place where a river has cut through the layer of pancakes and exposed them in cross section. In some places, the layers are buried. Elsewhere, they are bent or folded. And in other places, they have been completely eroded away. So when you look at a layer of rock, ask yourself three questions:

1. What do I see at the surface?
2. Where does that layer go underground?
3. (This is the tricky one.) If I look into the sky, where would the layer be that used to be there but isn't anymore, since it's eroded away?

So now you have a simple way to look at the rocks of the Bighorn Basin, where there are at least twenty-five distinct geologic formations. In places where they are all stacked up in a single pile, like the middle of the basin, they form a stack of layers that is more than 17,000 feet thick – that's more than three miles!

Opposite: Outcrop of the Meeteetse Formation near Meeteetse, Wyoming

Background: Fossil leaf from the Big Cedar Ridge site near Worland, Wyoming

The layers of rock contain all sorts of interesting features, including ripple marks, mud cracks, fossil root traces, and fossils. The nature of the rocks and what you find in them allows you to reconstruct what the landscape was like when the layer of sediment, which would eventually become rock, was being deposited. For this book, we used all kinds of information to conceive ideas of what these ancient Bighorn Basin landscapes used to look like, allowing the artist, Jan Vriesen, to bring them to life and make paintings of places that none of us had ever seen.

Drawing on the many years of our own research and teaching in the basin, we studied the rock layers at multiple locations around the Bighorn Basin. We read scientific papers written during the last 140 years about these rocks and their enclosed fossils. We visited museums and looked at the fossils in their collections. Altogether, we were able to assemble a group of ideas, sketches, and information that we gave to Jan, who composed a series of preliminary images. We then worked back and forth to create the paintings, which Jan did with acrylic paint on canvas. The results are a dozen visual hypotheses of what northwestern Wyoming looked like at twelve distinct times in the distant past.

Welcome to the Bighorn Basin

The northwestern corner of Wyoming is home to two of the nation's most famous national parks: Yellowstone and Grand Teton. Each year, these parks see more than 3 million visitors who arrive from all directions, and those who come to Yellowstone from the east by way of Cody must pass through a vast, dry depression known as the Bighorn Basin. The basin's main towns are Thermopolis, Cody, Powell, Lovell, Greybull, and Worland, but we'd be remiss if we didn't mention Shell, Ten Sleep, Meeteetse, Basin, Otto, and Bridger.

From outer space or on the Wyoming Highway Map, the basin appears as a giant oval hole about 150 miles long by 80 miles wide. The high points of the surrounding mountains reach more than 11,000 feet, while the low point of the basin is only 3,500 feet. The Bighorn Basin is a curious bit of topography, and it has one of the best geological stories on the planet.

In fact, the Bighorn Basin may be the best place on Earth to tell the story of our planet. Because of its geology, the Bighorn Basin contains layers of rock older than 2.5 billion years, as well as many, many younger rock layers. What makes this place so amazing is that it has layers of rock from almost every single geologic time period. If you had to pick one place in the world to tell the story of Earth's history, you would pick this place. So we picked this place.

The layered rocks of the Bighorn Basin were once ancient landscapes, and the fossils in the rocks are clues to what these landscapes looked like, what the ancient vegetation was, and what kinds of animals lived here. Because the Bighorn Basin is a dry place, not many plants grow here today, so it is easy to see the rocks. If you can see the rocks, you can find the fossils in the rocks. In this place, the history of the Earth lies on the ground as if it were an open book. And the goal of our little book is to give you the tools to read the big rock book of the Bighorn Basin.

Using layered rocks and fossils, geologists and paleontologists are able to envision what these lost worlds looked like. To share them with you, we studied the rocks; tracked down the fossils; reconstructed the plants, animals, and landscapes; and then employed an artist to paint them, choosing ancient worlds ranging in age from 520 million years to 18,000 years old. There are so many layers of rock in the basin that we could have painted hundreds of them. We chose to paint twelve.

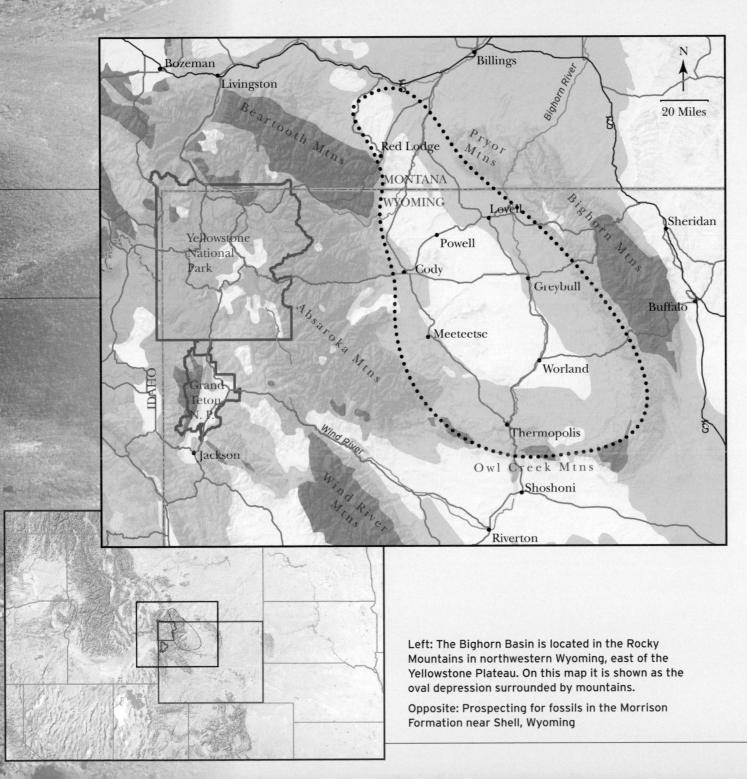

This geological map of northwestern Wyoming and adjacent Montana and Idaho shows the age of the rocks exposed at the surface. The colors represent:

■ **Cenozoic volcanic rock**
(66 million years ago through present day)

Cenozoic sedimentary rock
(66 million years ago through present day)

Mesozoic sedimentary rock
(252–66 million years ago)

Paleozoic sedimentary rock
(542–252 million years ago)

■ **Precambrian metamorphic rock**
(2.5–1.7 billion years ago)

Left: The Bighorn Basin is located in the Rocky Mountains in northwestern Wyoming, east of the Yellowstone Plateau. On this map it is shown as the oval depression surrounded by mountains.

Opposite: Prospecting for fossils in the Morrison Formation near Shell, Wyoming

AN EXTREMELY SHORT HISTORY
OF ROCKS, THE WORLD, AND THE BIGHORN BASIN

Our universe formed in the big bang around 13.8 billion years ago, and our solar system, with our sun and all of our planets, formed about 4.567 billion years ago. The early Earth was molten, but by 4 billion years ago, it had cooled enough that its surface was rocky, and water had filled large depressions to form the first oceans. Shortly thereafter, the first life in the form of bacteria appeared and began to colonize the planet.

The early continents were small and regularly broke apart and smashed into each other. The chunk that included what is now Wyoming formed about 2.5 billion years ago. By about 1.8 billion years ago, a number of small continents had collided to form the core of what would become North America. One geologist has called this the United Plates of America.

You can think of continents as thin, flat pieces of Styrofoam floating on a swimming pool — except the Styrofoam is hard rock, and the water is soft rock that can flow like super-gooey molasses. If you add a weight to the foam, it will float lower in the water. It is the same with continents: If you add weight to them, they will sink lower; if you remove weight, they will float higher. You can add weight by loading continents with sediments, ice sheets, or water. You can remove weight via erosion, melting ice sheets, or draining water.

When parts of a continent sink below sea level, the sea floods in and covers sections of it, and the weight of the water will make the continent sink more. Sediment carried into the sea from the continent's eroding mountains will add yet more weight and make the continent sink further, creating room for more sediment. As the sediment layers get buried by more layers, they are compacted and become rock. That is how layers of sedimentary rock form and get preserved underground for millions and millions of years.

The Bighorn Basin is one of those singular places on Earth where lots of layers were deposited and stored underground for millions of years. They were later pushed up and exposed at the surface so that today we can see them in all of their glory. These layered rocks that poke out of the ground around the basin tell a pretty amazing story about our planet.

Although the first 3 billion years of life on Earth were ultimately dominated by very simple single-celled organisms like bacteria, by 542 million years ago, a host of more complicated multicelled marine animals had evolved. As the sediment piled up in places like the Bighorn Basin, it also piled up the bodies of the dead sea animals. The planet buries its dead, and the buried dead are called fossils.

Opposite: Cottonwoods in fall in Wyoming

Inset: Abeba Ender struggles to hold *Wilsonoceras bighornense*, a large coiled cephalopod from the Ordovician Bighorn Dolomite.

Right: The view to the east from the flanks of the Beartooth Mountains near Red Lodge, Montana. The fins of rock near the girl's elbow are beds of the Bighorn Dolomite that have raised over time so the once-horizontal layers are now almost vertical.

Between 542 and 60 million years ago, the northwestern part of Wyoming was on a part of North America that was sinking and accumulating layers of sediment that would become fossil-filled rock layers. About 60 million years ago, the mountains that today ring the Bighorn Basin began to form. As they formed, they lifted up and folded the rock layers that were already there. As those rock layers rose and became exposed at the surface, they also eroded, and the sand and mud from the mountains flowed down into the basin and began to fill it up.

During this time, there were some pretty aggressive volcanoes in westernmost Wyoming. These mountains erupted repeatedly between 50 and 45 million years ago, forming what are now called the Absaroka Mountains.

By 45 million years ago, the Bighorn Basin was fully ringed by mountains. The rivers flowed off of those mountains and into the basin, and continued to fill it with sediment. Eventually the entire basin was full, and the surrounding mountains were buried in their own eroded debris.

Then a very strange thing happened. The whole region began to lift up and rivers began to cut down through both the mountains and the basin. Eventually these rivers carved the Wind River, Bighorn, and Clarks Fork Canyons through the margins of the basin. It was through these outlets that the eroding rivers scoured the basin, exposing the rock layers and creating the topography we see today.

It is in these canyons and in these rock layers that this story is told.

How Rock Layers Relate to Geologic Time

Here we connect the geologic episodes that we discuss in the book to specific rock formations, and the geological age in which they formed. The whole pile of rock layers in the Bighorn Basin (right) represents about 540 million years of time, and each individual formation represents a smaller window within that time.

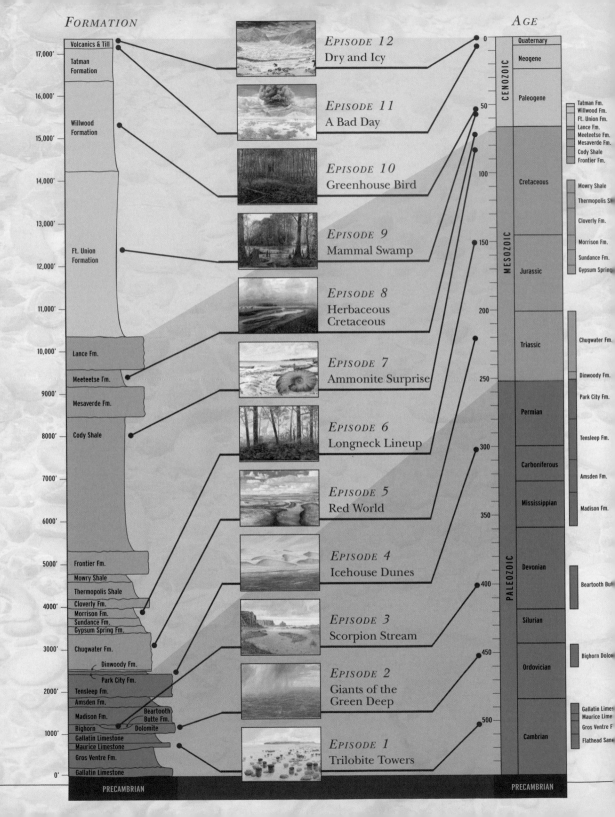

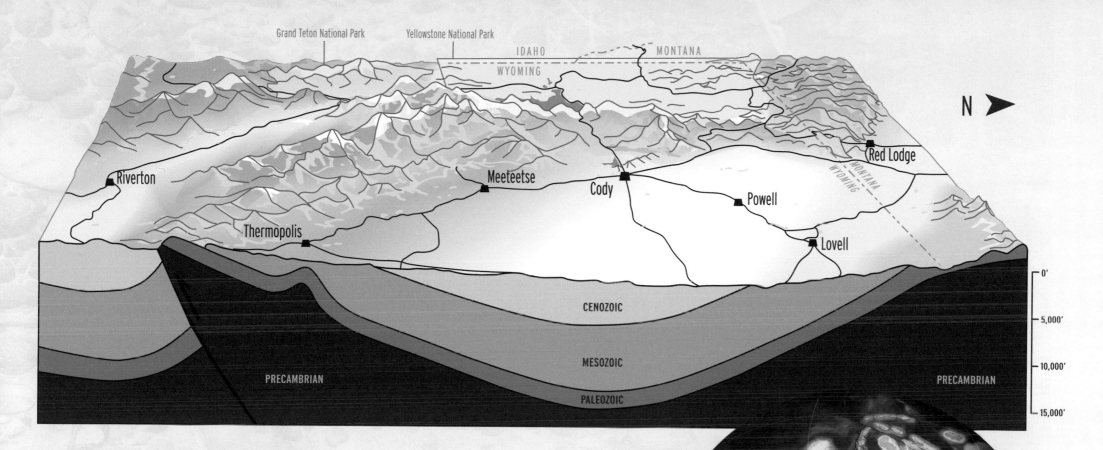

Grand Teton National Park

Yellowstone National Park

IDAHO MONTANA

WYOMING

N ▶

Riverton

Meeteetse

Cody

Powell

Red Lodge

WYOMING
MONTANA

Thermopolis

Lovell

CENOZOIC

— 0'

— 5,000'

MESOZOIC

— 10,000'

PRECAMBRIAN

PALEOZOIC

PRECAMBRIAN

— 15,000'

Bighorn Basin Cross Section

This block diagram shows the Bighorn Basin as if it were sliced down the middle from north (right) to south (left), approximately along the Bighorn River. This perspective shows how the geological formations extend underground from what you can see exposed at the surface.

At the deepest part the basin, the entire thickness of sedimentary rocks is about 17,000 feet (see the diagram on page 10 for details). On the left side of the diagram, you can also see the northern edge of the Wind River Basin, which is just to the south of the Bighorn Basin.

Maps of Ancient Earth

Because the Earth's continents are always moving, different time periods have different maps. These paleogeographic maps show what the land that is now North America looked like and where what is now Wyoming could be found during each of the episodes we describe in this book. In these maps, the dark blue sections are ocean, the light blue is shallow sea, and the solid line is the equator. The red rectangle is Wyoming.

Above: University of Wyoming geologists in the field in southern Wyoming, circa 1930

Inset: Paleobotanist Scott Wing holds a section of core drilled from the Fort Union Formation near Basin, Wyoming.

The Bighorn Basin Coring Project

Geologists and paleontologists have been coming to the Bighorn Basin since the 1880s, and they keep coming back because there is so much amazing geology to see and so many incredible fossils to find. Many of them have realized that the geology has direct value in the form of oil, gas, coal, gypsum, bentonite, and groundwater. Others come because they realize that the Bighorn Basin really is one of the best places in the world to study specific events in Earth's long history.

In 2011, a team of forty earth scientists from around the world came to the Bighorn Basin to drill and core three wells in the Willwood Formation. They came here because it is probably the best place on the planet to study what happened on land 55.5 million years ago, when there was a sudden and large burst of global warming known as the Paleocene-Eocene Thermal Maximum. Sometimes things that happened millions of years ago can have great relevance to planet Earth today.

Left: A section of the core of the Willwood Formation, showing red mudstone and small nodules of limestone.

Right: The Bighorn Basin Drilling Project worked with an experienced team of drillers to extract 3,000 feet of core.

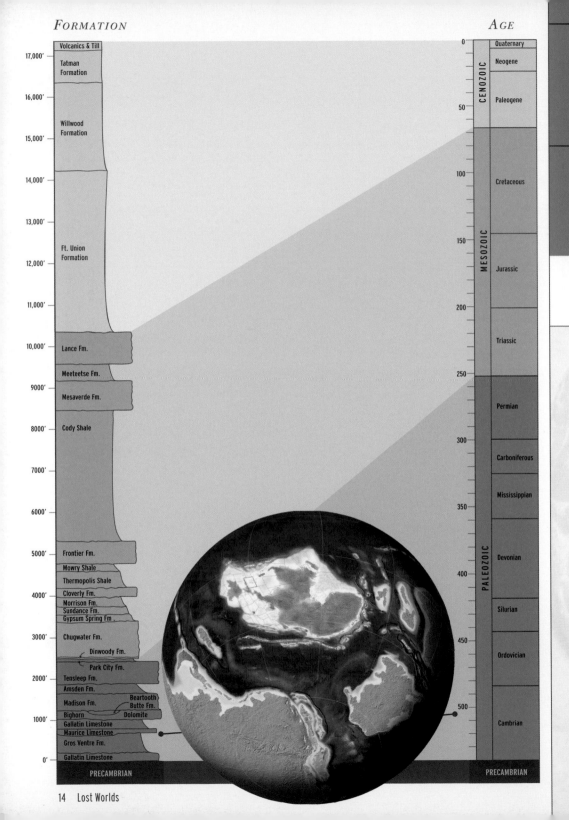

	Formation		Age	
17,000'	Volcanics & Till		0	Quaternary
	Tatman Formation	CENOZOIC		Neogene
16,000'			50	Paleogene
15,000'	Willwood Formation			
14,000'			100	Cretaceous
13,000'		MESOZOIC		
12,000'	Ft. Union Formation		150	Jurassic
11,000'			200	
10,000'	Lance Fm.			Triassic
9000'	Meeteetse Fm.		250	
8000'	Mesaverde Fm.			Permian
	Cody Shale		300	
7000'				Carboniferous
6000'			350	Mississippian
5000'	Frontier Fm.	PALEOZOIC		Devonian
4000'	Mowry Shale / Thermopolis Shale / Cloverly Fm. / Morrison Fm. / Sundance Fm. / Gypsum Spring Fm.		400	Silurian
3000'	Chugwater Fm. / Dinwoody Fm.		450	Ordovician
2000'	Park City Fm. / Tensleep Fm. / Amsden Fm.			
1000'	Madison Fm. / Bighorn / Beartooth Butte Fm. / Dolomite / Gallatin Limestone / Maurice Limestone / Gros Ventre Fm.		500	Cambrian
0'	Gallatin Limestone			
	PRECAMBRIAN			PRECAMBRIAN

Lost Worlds

Trilobite Towers

EPISODE	AGE:	FORMATION:	ANCIENT ENVIRONMENT:
1	500 MILLION YEARS, CAMBRIAN	MAURICE LIMESTONE	WARM AND WET

Past

Mounds of slimy algae and photosynthetic bacteria poke up above a muddy bottom in warm, shallow seas. Rocky ridges in the distance are barren of plants and animals – nothing growing, nothing walking, nothing moving, just a thin veneer of soil and lots of rock. In the water, small insect-like trilobites scuttle between the algae towers, and filter-feeding lampshell brachiopods cling to the bottom. The ocean here is that light-blue color that occurs near the equator, and the water leaves a salt stain when it evaporates from a rock. The scene is calm and peaceful, but a biological revolution is under way – a revolution that will forever change evolution of life on Earth and our ability to study it using the fossil record.

What you see today

One of the oldest sedimentary layers on top of the meta-morphic and igneous Precambrian "basement" rocks is the Maurice Limestone, which was deposited during the Cambrian Period. Because the oldest layers in a basin are on the edges near the adjacent mountains, that is where you can see the Maurice. The pictured outcrop can be found in the Clarks Fork Canyon west of Clark, Wyoming. Here, the Clarks Fork River tumbles out of the Beartooth Mountains, cutting through the entire sequence of tilted geological layers and exposing them on the sides of the canyon. The Maurice Formation tends to form a cliff because it is hard, well-cemented lime-stone surrounded by softer, weakly cemented mud-stone and shale.

Left: A *Modocia typicalis* trilobite from Utah. This same animal has been found in the Bighorn Basin.

Top right: A slab of bedrock that is made of lots of small, flat pebbles

Bottom right: Outcrops of the Maurice Formation in Clarks Fork Canyon near Clark, Wyoming

Significance

During the Cambrian, Wyoming formed the coastline of North America and was situated near the equator. The present-day land to the west (Idaho, Nevada, Oregon, Washington, and California) had not yet crashed into the North American continent, so there was little geological drama nearby in the form of mountain building or volcanoes. Sea level was rising around the globe, and the Bighorn Basin represented a calm, shallow marine environment. Multicelled organisms with hard parts, such as trilobites and brachiopods, were beginning to rapidly evolve, but simpler single-celled holdovers from the Precambrian, including mats of bacteria and algae called stromatolites, were still common.

This was the time of the greatest revolution in the history of life on Earth – the "Cambrian Explosion." During the approximately 3.5 billion years before the Cambrian, the biosphere was dominated by single-celled organisms. The first hints of multicelled life actually appeared just before the beginning of the Cambrian, but these life-forms had no shells or skeletons to speak of and thus were rarely fossilized. Then, multicelled organisms began to secrete hard parts, and within a few million years the fossil revolution had begun. When organisms started growing shells and skeletons, they essentially made their own rock, which means they more easily became fossils for us to find millions of years later (or hundreds of millions of years in this case!). While the shells and skeletons provide the animals with anatomical structure and protection from predators, they provide us with a much clearer window into the biological history of our planet than is available for the soft-bodied world of the Precambrian.

Upper right: In Cambrian coastal lagoons, layers of microbes made a thin film that coated the seafloor like sticky glue. Occasionally, storms would scour the seafloor and rip up layers of mud that were held together by the microbes. When the water calmed, the flat flakes were left as piles on the seafloor, making a flat pebble conglomerate. This rock type became extinct once the animals were common enough to graze away the bacterial film.

Inset: Hiking toward the outcrop in Clarks Fork Canyon

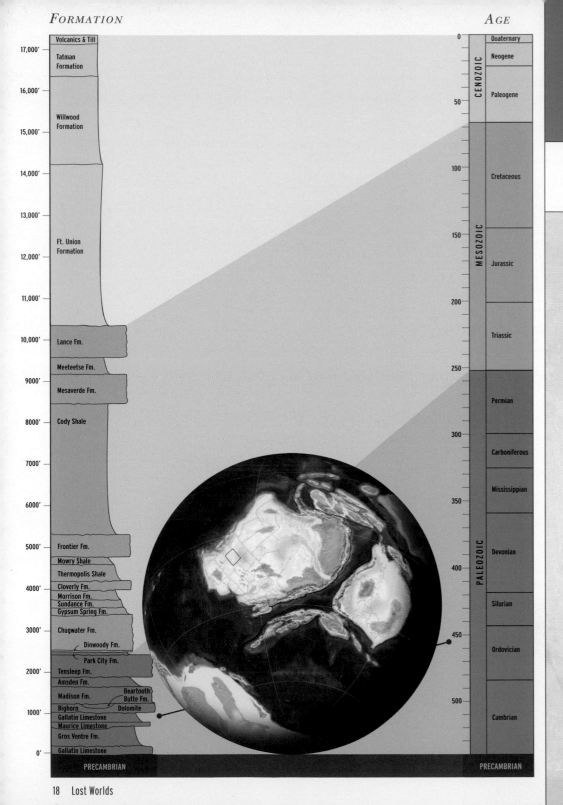

FORMATION

17,000'	Volcanics & Till
	Tatman Formation
16,000'	
15,000'	Willwood Formation
14,000'	
13,000'	Ft. Union Formation
12,000'	
11,000'	
10,000'	Lance Fm.
9000'	Meeteetse Fm.
	Mesaverde Fm.
8000'	Cody Shale
7000'	
6000'	
5000'	Frontier Fm.
	Mowry Shale
4000'	Thermopolis Shale
	Cloverly Fm.
	Morrison Fm.
	Sundance Fm.
	Gypsum Spring Fm.
3000'	Chugwater Fm.
	Dinwoody Fm.
	Park City Fm.
2000'	Tensleep Fm.
	Amsden Fm.
	Madison Fm. / Beartooth Butte Fm.
1000'	Bighorn Dolomite
	Gallatin Limestone
	Maurice Limestone
	Gros Ventre Fm.
0'	Gallatin Limestone
	PRECAMBRIAN

AGE

0	Quaternary
	Neogene
50	CENOZOIC — Paleogene
100	Cretaceous
150	MESOZOIC
200	Jurassic
	Triassic
250	Permian
300	Carboniferous
350	Mississippian
400	PALEOZOIC — Devonian
	Silurian
450	Ordovician
500	Cambrian
	PRECAMBRIAN

Giants of the Green Deep

EPISODE	AGE:	FORMATION:	ANCIENT ENVIRONMENT:
2	450 MILLION YEARS ORDOVICIAN	BIGHORN DOLOMITE	WARM AND WET

Past

A broad, shallow tropical sea stretches as far as the eye can see. A pod of giant squid hovers near the surface. Some of these squid have huge cylindrical shells that are more than a foot in diameter and up to 30 feet long. They look like floating logs, but they lived long before the first tree. Like today's squid, they are hunters, and they are the largest predators of their time. Gathering together near the surface to school and spawn, they have the sea largely to themselves; there are almost no fish and no other large animals. Stretching across the limy seafloor far below, shelly sea creatures go about their business. Smaller shelled squid, flower-like crinoids, feathery bryozoans, cone-shaped corals, clam-like brachiopods, bug-like trilobites, and coiled snails live in a variety of communities and are the source of food for the giant squid.

What you see today

Today, the Bighorn Dolomite forms one of the more recognizable geologic layers in the mountains that surround the Bighorn Basin. It forms prominent creamy vertical cliffs that rise above the slopes formed by the older underlying shale. When these cliffs fail, they send massive rectangular blocks smashing down to the bottom of the valley. The blocks look like giant dice, and one of them sits right in the middle of the Wind River, six miles north of the Boysen Dam. Up close, the Bighorn Dolomite has a jagged texture, and geologists call it the "tear pants" formation because that's exactly what it does if you sit on it in the wrong place. Fossils are not common in this layer, but there are enough to tell the story of how these towering cliffs were made from what was once the bottom of the sea.

Left: The Bighorn Dolomite forms the prominent cliff on the wall of the Wind River Canyon south of Thermopolis, Wyoming. Large blocks of this formation litter the slope below the cliff and one has tumbled into the middle of the Wind River.

Significance

The Bighorn Dolomite formed as limy mud at the bottom of a shallow sea. At the time, North America was far south of its present position, and the state of Wyoming would have been about 10 degrees south of the equator. The animals that lived in this sea were diverse, representing early examples of many groups of organisms that would continue to dominate life in the salty seas for hundreds of millions of years.

Top: Torianne, Skip, and Alexis Hommer can barely lift this eight-foot segment of a giant *Endoceras* cephalopod fossil. The sketch behind the photo shows what the animal would have looked like when it was alive. These tentacled animals were related to modern squid and octopus, and lived in the Ordovician sea.

Right: Fossil chain coral, *Halysites gracilis*, from the Bighorn Dolomite. These fossils, which can be found high in the mountains of Wyoming, are direct evidence of ancient seas that covered portions of North America.

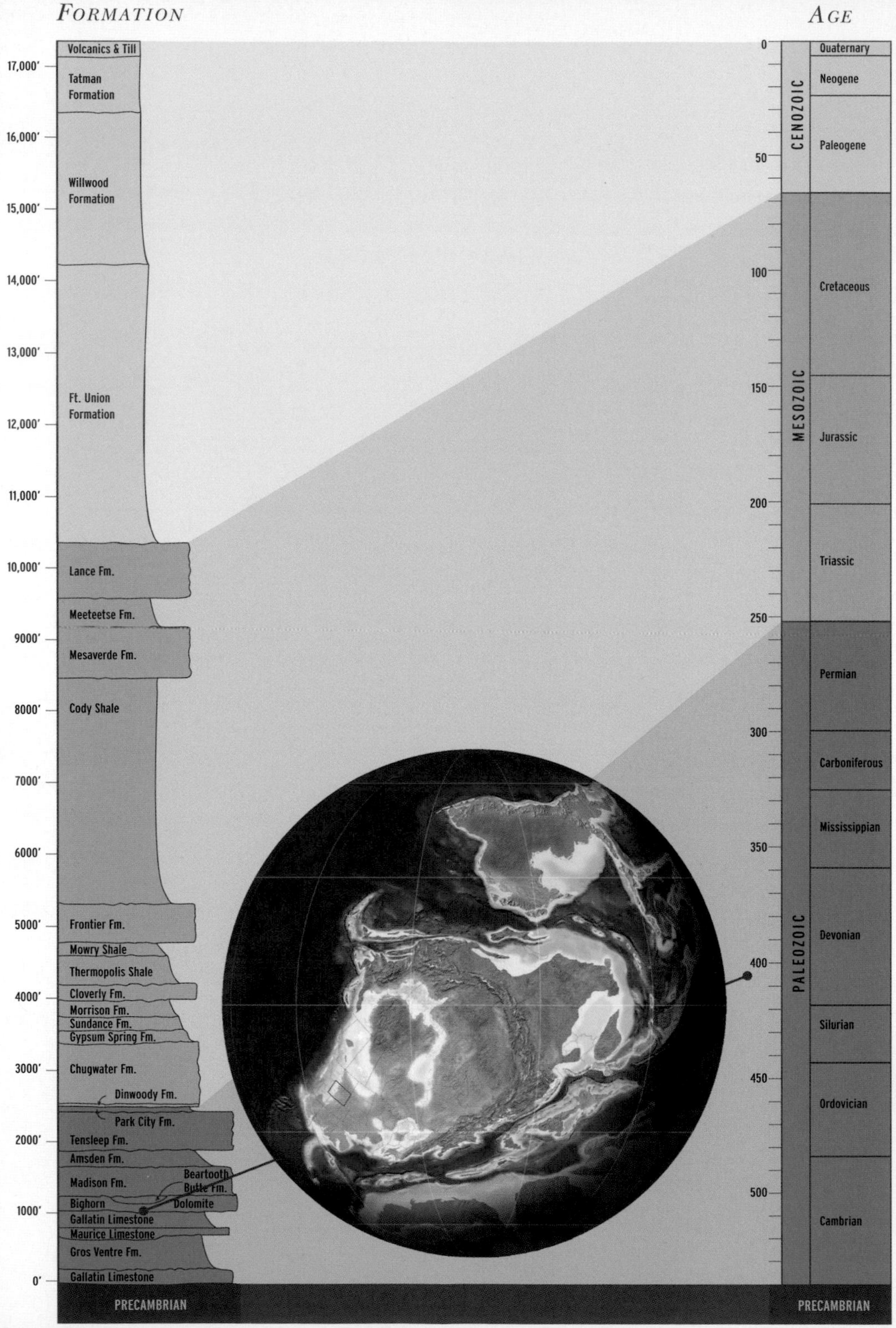
FORMATION
AGE
Volcanics & Till
Tatman Formation
Willwood Formation
Ft. Union Formation
Lance Fm.
Meeteetse Fm.
Mesaverde Fm.
Cody Shale
Frontier Fm.
Mowry Shale
Thermopolis Shale
Cloverly Fm.
Morrison Fm.
Sundance Fm.
Gypsum Spring Fm.
Chugwater Fm.
Dinwoody Fm.
Park City Fm.
Tensleep Fm.
Amsden Fm.
Madison Fm.
Beartooth Butte Fm.
Bighorn Dolomite
Gallatin Limestone
Maurice Limestone
Gros Ventre Fm.
Gallatin Limestone
PRECAMBRIAN
17,000'
16,000'
15,000'
14,000'
13,000'
12,000'
11,000'
10,000'
9000'
8000'
7000'
6000'
5000'
4000'
3000'
2000'
1000'
0'
0
50
100
150
200
250
300
350
400
450
500
CENOZOIC
MESOZOIC
PALEOZOIC
Quaternary
Neogene
Paleogene
Cretaceous
Jurassic
Triassic
Permian
Carboniferous
Mississippian
Devonian
Silurian
Ordovician
Cambrian
PRECAMBRIAN

Scorpion Stream

EPISODE	AGE:	FORMATION:	ANCIENT ENVIRONMENT:
3	400 MILLION YEARS, DEVONIAN	BEARTOOTH BUTTE FORMATION	WARM AND DRY

Past

Streams are entering a coastal area and have cut into the surrounding bedrock of Bighorn Dolomite. The channels are filling with sediment that has eroded from the surrounding hills. Lurking under the brackish water are armored fish, snails, and brachiopods. A five-foot-long predatory eurypterid is trolling the shallows in search of its next meal. These "scorpions of the water" are some of the largest predators of the Paleozoic and close evolutionary cousins to spiders and horseshoe crabs. They have legs for walking and paddles for swimming, so they can easily move in and out of the water. On land, life is now apparent. Thin, low-stemmed plants are sprouting from the muddy deposits on the edges of the streams. True land scorpions are scurrying among the plants, hunting for other critters that have evolved into this new, wide-open ecosystem outside of the water.

What you see today

The most spectacular outcrop of the Beartooth Butte Formation is perched on the top of the Beartooth Plateau, more than 6,000 feet above the basin floor. This geological remnant is the only bit of post-Precambrian sedimentary rock left on top of the mountains in this area – the rest was eroded away during the rise of the Rockies. The butte preserves horizontal layers of Cambrian, Ordovician, and Devonian shale and limestone, with sediments of the Beartooth Butte Formation filling channels cut down into the Bighorn Dolomite. These channels formed when sea level dropped during the Early Devonian, creating a coastal environment where streams flowed in from the adjacent land. Sediment slowly filled these channels, entombing the pieces and parts of the organisms that were living in this thriving ecosystem.

Left: The claw of a sea scorpion known as a *Pterygotus*. This claw is nearly six inches long, and the animal that owned it was more than five feet long.

Above: Rock slabs containing pieces of the bony head plates of armored fish called placoderms.

Below: Beartooth Butte near Cooke City, Montana, is the best place to see the Beartooth Butte Formation, which is the mass of red rock in the face of the butte.

Significance

The beginning of the Devonian was the time when organisms were just emerging onto land. The early land plants were small – no forests existed yet, just low stems and a few small leaves. Plant roots and dead plant debris mixed in with weathered rock to form soils that started to live and breathe like the ones we have today. Arthropods, the evolutionary group that includes crabs, insects, and trilobites, were the first animals to be preserved as fossils from this new land-based ecosystem, but other soft-bodied groups were likely there also, leaving behind evidence in the form of burrows and tracks. The Beartooth Butte Formation contains a mix of marine (brachiopods and snails) and land (scorpions and plants) organisms, providing a perfect window into the very environment where this remarkable water-to-land evolutionary transition was taking place.

Upper left: A red layer of the Beartooth Butte Formation can be seen in Cottonwood Canyon in the Bighorn Mountains near Lovell, Wyoming.

Upper right: The view south from Beartooth Butte toward the Absaroka Mountains

Lower left: A rock slab with multiple head plates of armored fish called placoderms

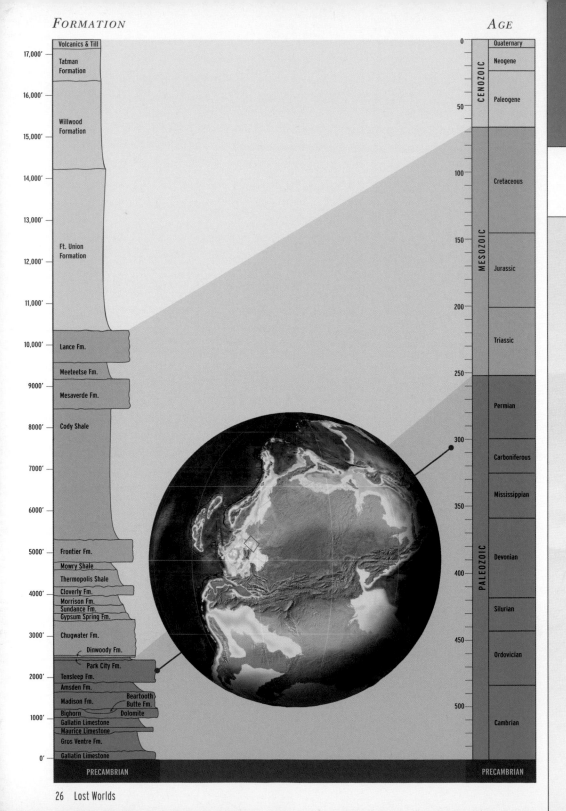

FORMATION

17,000' — Volcanics & Till
Tatman Formation
16,000'
Willwood Formation
15,000'
14,000'
Ft. Union Formation
13,000'
12,000'
11,000'
10,000' — Lance Fm.
Meeteetse Fm.
9000' — Mesaverde Fm.
8000' — Cody Shale
7000'
6000'
5000' — Frontier Fm.
Mowry Shale
Thermopolis Shale
4000' — Cloverly Fm.
Morrison Fm.
Sundance Fm.
Gypsum Spring Fm.
3000' — Chugwater Fm.
Dinwoody Fm.
Park City Fm.
2000' — Tensleep Fm.
Amsden Fm.
Madison Fm. — Beartooth Butte Fm. Dolomite
Bighorn
Gallatin Limestone
1000' — Maurice Limestone
Gros Ventre Fm.
0' — Gallatin Limestone

PRECAMBRIAN

AGE

0 — Quaternary
Neogene
CENOZOIC
50 — Paleogene
100 — Cretaceous
MESOZOIC
150 — Jurassic
200 —
Triassic
250 —
Permian
300 — Carboniferous
Mississippian
350 —
PALEOZOIC
Devonian
400 —
Silurian
450 — Ordovician
500 — Cambrian

PRECAMBRIAN

Icehouse Dunes

EPISODE	AGE:	FORMATION:	ANCIENT ENVIRONMENT:
4	300 MILLION YEARS, PENNSYLVANIAN	TENSLEEP SANDSTONE	COOL AND DRY

Past

Wind-whipped waves roll ashore on a desert coastline. Mile after mile of giant 40-foot-high sand dunes stretch away from the beach. As far as the eye can see, there is nothing but water and sand, and while the shallow sea is warm and tropical, there are no reefs. The wind blows sand grains up the shallow upwind slope of the dunes and deposits them on the steeper downwind side. In this way the dunes slowly creep across the landscape.

What you see today

The Tensleep sandstone is very distinctive because the internal anatomy of the ancient fossilized sand dunes is clearly visible. In places where the formation itself is flat, the internal layers are tilted 15 degrees or so off horizontal. These are the slopes of the dunes, now buried and hardened. Layers like this clearly show the wind direction when the rocks were just sand. In some places, the layer contains marine fossils such as brachiopods, snails, crinoids, and even trilobites. This is indisputable evidence that these dunes were once coastal.

Left: A fossil brachiopod (balanced on a fingertip) shows that parts of the Tensleep Formation were deposited under shallow salt water.

Above: Inclined layers of sand in a modern-day riverbank show how currents can cause "cross-beds." These sloping layers of sand can be preserved in ancient sandstone and are useful for understanding ancient environments. Very large cross-beds are typically found in sand dunes.

Significance

By this time, the great supercontinent of Pangaea had formed, and the giant landmass greatly affected the climate, creating large swaths of arid land and the origin of large deserts. During the time of Pangaea, huge ice caps formed near the South Pole. There was no land at the North Pole so, unlike the Pleistocene ice ages, this one had an ice cap on the South Pole but not the North. Over much of western North America, parts of this ice age were expressed as cold, dry climates that reduced vegetation and resulted in extensive sandy deserts.

Upper left: Cross-beds preserved in the Tensleep Sandstone are evidence of ancient sand dunes.

Below: In this outcrop near Shell, Wyoming, the steeply dipping cross-beds of the Tensleep Sandstone are overlain by the gently dipping layers of the Park City Formation.

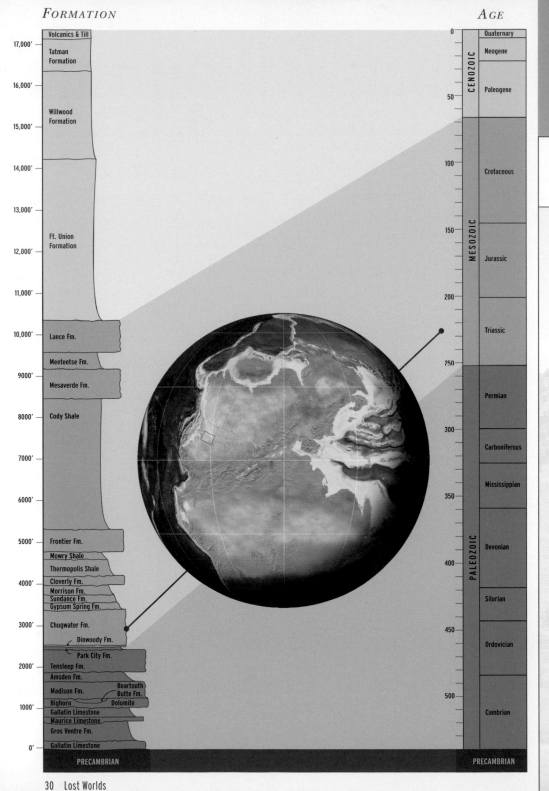

17,000'	Volcanics & Till
	Tatman Formation
16,000'	
	Willwood Formation
15,000'	
14,000'	
13,000'	
	Ft. Union Formation
12,000'	
11,000'	
10,000'	Lance Fm.
9,000'	Meeteetse Fm.
	Mesaverde Fm.
8,000'	Cody Shale
7,000'	
6,000'	
5,000'	Frontier Fm.
	Mowry Shale
	Thermopolis Shale
4,000'	Cloverly Fm.
	Morrison Fm.
	Sundance Fm.
	Gypsum Spring Fm.
3,000'	Chugwater Fm.
	Dinwoody Fm.
	Park City Fm.
2,000'	Tensleep Fm.
	Amsden Fm.
	Madison Fm. / Beartooth Butte Fm.
1,000'	Bighorn Dolomite
	Gallatin Limestone
	Maurice Limestone
	Gros Ventre Fm.
0'	Gallatin Limestone
	PRECAMBRIAN

Age scale:
- 0 — Quaternary (CENOZOIC)
- Neogene
- 50 — Paleogene
- 100 — Cretaceous (MESOZOIC)
- 150 — Jurassic
- 200 — Triassic
- 250 — Permian (PALEOZOIC)
- 300 — Carboniferous
- Mississippian
- 350 — Devonian
- 400 — Silurian
- 450 — Ordovician
- 500 — Cambrian
- PRECAMBRIAN

Red World

EPISODE	AGE:	FORMATION:	ANCIENT ENVIRONMENT:
5	220 MILLION YEARS, TRIASSIC	CHUGWATER FORMATION	HOT AND SEASONALLY DRY

Past

Deeply colored red mudflats can be seen far into the distance. Shallow channels drain the area, and little life is evident. Strong thunderstorms are booming away in the distance over a faraway forest. A lone rhynchosaur ambles across the plain, leaving a track in the soft mud underneath. This creature is like a vertebrate mash-up: a wide head with a short snout that resembles a hammerhead shark's, a stout parrot-like beak, and plates of bumpy fish-like teeth lining its mouth so it can grind the plants that make up its diet. Sharp claws on its back feet could be used to dig up roots to eat or to protect it from the vast array of crocodile-like predators that roam the landscape.

What you see today

The Triassic Chugwater Formation is the most recognizable geological unit in the basin. Its bright-red color makes it stand out among the other more subdued colors of the adjacent formations. In fact, these red rocks can be seen clearly as you fly over the basin in an airplane and even in satellite imagery from space. During the Triassic, Wyoming was in the northern tropics, and North America was starting to grow to the west by colliding with smaller landmasses. The Chugwater, like the other Paleozoic and older Mesozoic units in the basin, is usually found along the basin margin, folded up during the subsequent rise of the Rockies. Because of this, the Chugwater forms a red ring around most of the basin when seen from above.

Left: At the mouth of Clarks Fork Canyon, the Chugwater Formation has been folded by the uplift of the Beartooth Mountains.

Below: The Chugwater Formation south of Ten Sleep, Wyoming

Significance

The Chugwater's red color is very common for rocks of this age all over the world. It is rust, an oxidized form of iron that is also known as the mineral hematite. Just like a nail rusts when it is exposed to moisture and allowed to dry, sediments rust, and get red, when they experience cycles of wetting and drying. Red sediments are common today in places that have strong seasonal changes in rainfall, like the tropical and continental interior areas that experience monsoons. Why so much rust in the Triassic? This is when all of the world's continents had come together into the great supercontinent called Pangaea. Today, the largest monsoons occur on the largest continents, which means a supercontinent like Pangaea likely had a "mega-monsoon." These extreme wet and dry seasons during the Triassic caused massive rusting of the sediments, leaving behind a geological red ribbon that can be seen on all of the continents. The rusting process in the sediments often destroys the remains of plants and animals that would otherwise become fossilized, so very few fossils have been discovered in the Chugwater over the years.

Top left: The footprint of a Triassic reptile (*Chirotherium barthii*) preserved in a slab of red sandstone. The track is roughly the size of a human hand.

Middle: Fossils are extremely rare in the Chugwater Formation, so this tooth of a Triassic reptile is quite a find.

Below: In places, the Chugwater Formation has been tilted so that once-horizontal beds are now vertical.

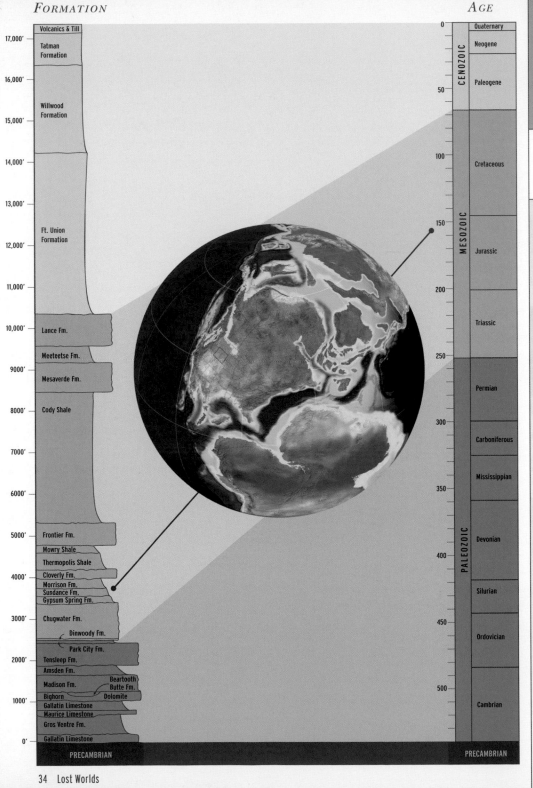

Longneck Lineup

EPISODE	AGE:	FORMATION:	ANCIENT ENVIRONMENT:
6	150 MILLION YEARS, JURASSIC	MORRISON FORMATION	WARM AND WET

Past

It is a foggy, quiet morning at the edge of a dense forest. The trees look vaguely familiar but on second glance are clearly not. In the distance, barely visible is a group of huge dinosaurs with long necks and tiny heads. They are moving very slowly and deliberately as they feed their way across a meadow of ferns and horsetails. There is no menace, only the muffled footfalls of massive herbivores.

What you see today

The Morrison Formation was first described in Colorado, and rocks of this name stretch across Utah and up through Wyoming. In the Bighorn Basin, the formation is very colorful with hues of blue, red, orange, and brown, but it is famous for its immense and diverse dinosaurs. Because of its high clay content, the formation does not make prominent outcrops, and it is often covered by geological debris or vegetation. Like all of the Paleozoic and Mesozoic formations in the basin, the Morrison crops out around the basin's rim. Most of the best dinosaur fossils have come from the eastern edge, and active dinosaur quarries occur from Thermopolis to Shell.

In 1934, Barnum Brown from the American Museum of Natural History (AMNH) in New York opened the Howe Quarry east of Greybull, Wyoming. This expedition was funded by Sinclair Oil and resulted in the company's green dinosaur symbol. The Howe Quarry yielded a remarkable dinosaur bone bed, including the *Barosaurus* skeleton that now stands on its hind legs in the Theodore Roosevelt atrium of the AMNH. In 1991, a 95 percent complete *Allosaurus* skeleton, now at the Museum of the Rockies in Bozeman, Montana, was collected at the Howe Quarry. The quarry also preserves carbonized remains of large trees and cones of extinct conifers.

Left: Barnum Brown at Howe Quarry, Shell, Wyoming, in 1934

Right: Fossilized fern leaves show that the climate was warm and wet 150 million years ago.

Above: A distant view of the Morrison Formation

Right: An elegant skeleton of a young long-necked *Diplodocus* dinosaur from a quarry near Shell, Wyoming

Significance

The Morrison Formation is perhaps the best window into the world of giant Jurassic dinosaurs, but the picture is cloudy because bones are much more frequently preserved than are plants. The result is a world where we know the animals but are only now beginning to understand the nature of the vegetation. This is all the more important because the long-necked sauropods, who were clearly herbivores, are the largest animals ever to walk the Earth, yet we have little real understanding about what they ate. Sites like the Howe Quarry are beginning to change that.

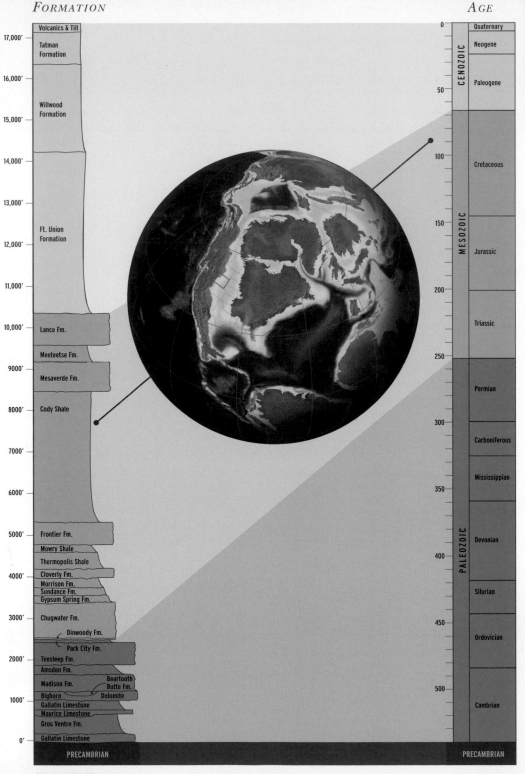

Quaternary

CENOZOIC

Neogene

Paleogene

Cretaceous

MESOZOIC

Jurassic

Triassic

Permian

Carboniferous

Mississippian

Devonian

PALEOZOIC

Silurian

Ordovician

Cambrian

PRECAMBRIAN

17,000'
Volcanics & Till
Tatman Formation
16,000'
Willwood Formation
15,000'
14,000'
13,000'
Ft. Union Formation
12,000'
11,000'
10,000'
Lance Fm.
9000'
Meeteetse Fm.
Mesaverde Fm.
8000'
Cody Shale
7000'
6000'
5000'
Frontier Fm.
Mowry Shale
Thermopolis Shale
4000'
Cloverly Fm.
Morrison Fm.
Sundance Fm.
Gypsum Spring Fm.
3000'
Chugwater Fm.
Dinwoody Fm.
Park City Fm.
2000'
Tensleep Fm.
Amsden Fm.
Beartooth Butte Fm.
Madison Fm.
1000'
Bighorn Dolomite
Gallatin Limestone
Maurice Limestone
Gros Ventre Fm.
0'
Gallatin Limestone

0
50
100
150
200
250
300
350
400
450
500

Ammonite Surprise

EPISODE	AGE:	FORMATION:	ANCIENT ENVIRONMENT:
7	**82 MILLION YEARS, CRETACEOUS**	**CODY SHALE** (TELEGRAPH CREEK MEMBER)	**WARM AND WET**

Past

The shell of a giant, six-foot-wide, coiled, squid-like ammonite lies on a beach partially buried by the sand. It is surrounded by other, much smaller shells of its evolutionary siblings. Nearby, a lobster has washed up, and sand is starting to bury a frond that has broken off of a nearby palm tree. Driftwood is scattered across the area, and flying reptiles known as pteranodons soar overhead, looking to dive down for their next strike. In the distance is a warm, shallow sea where a variety of giant marine reptiles hunt for one of their favorite meals: the ammonites.

What you see today

The Telegraph Creek Member is a sandy beach unit lying at the top of the distinctive black mudrocks of the Cody Shale – the black shales representing deeper water conditions than the overlying sandstones. Ground-up debris of dead plankton and dead plants that drifted offshore and became buried with the settling mud – all before it could be eaten by marine organisms or broken down by microbes – provides the black color. As the sea level dropped, the Telegraph Creek sands were deposited on the shore, and ammonite and other marine animal shells washed up, where they were buried by the moving sand.

Above: This ammonite, *Scaphites depressus*, is a common species in the marine rocks found in the Cody Shale. In life, this shell would have contained a squid-like ocean predator.

Right: A piece of a fossilized palm frond from North Dakota is quite similar to those that can be found in the rocks at the top of the Cody Shale. Its presence signals a warm and pleasant climate.

Significance

Global temperatures during the middle part of the Cretaceous were at an all-time high. This greenhouse world also witnessed some of the highest known sea levels, causing the interiors of continents to be inundated with shallow seaways. The warm, shallow waters of the North American "western interior seaway" formed a kind of marine sanctuary, home to a wide variety of animals including snails, giant clams, and many species of fish and ammonites. Some of the best fossils of marine reptiles come from these deposits. The high diversity of life and steady sediment supply from adjacent landmasses combined to make these continental seaways an excellent environment for fossil preservation.

Above right: The tiny ammonite *Scaphites hippocrepis* never grew to be much larger than a quarter, but it shared the Cody sea with marine reptiles and giant, six-foot ammonites.

Left: It is easy to recognize outcrops of Cretaceous marine shale because they weather to dark-gray slopes and valleys. Their fine layers are only clearly visible when they have been freshly cut by road construction or rivers.

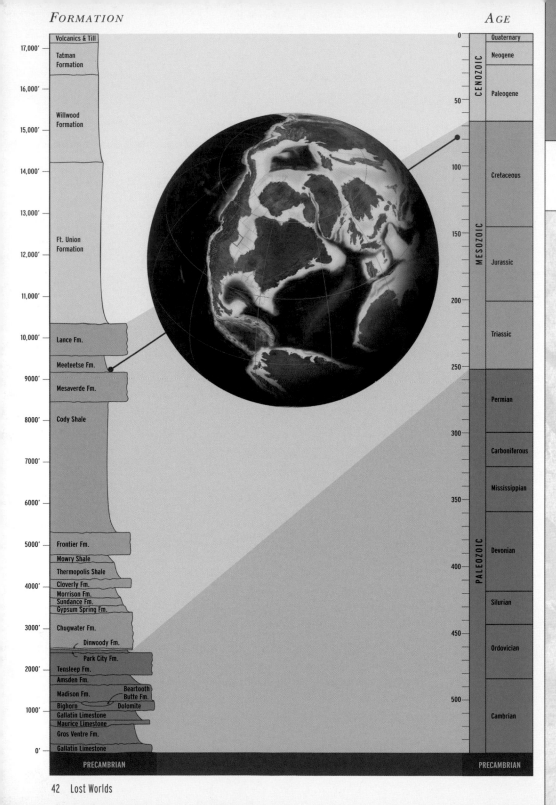

17,000'	Volcanics & Till
	Tatman Formation
16,000'	
	Willwood Formation
15,000'	
14,000'	
13,000'	
	Ft. Union Formation
12,000'	
11,000'	
10,000'	Lance Fm.
9000'	Meeteetse Fm.
	Mesaverde Fm.
8000'	Cody Shale
7000'	
6000'	
5000'	Frontier Fm.
	Mowry Shale
	Thermopolis Shale
4000'	Cloverly Fm. / Morrison Fm. / Sundance Fm. / Gypsum Spring Fm.
3000'	Chugwater Fm.
	Dinwoody Fm.
	Park City Fm.
2000'	Tensleep Fm.
	Amsden Fm.
	Madison Fm. / Beartooth Butte Fm.
1000'	Bighorn / Dolomite
	Gallatin Limestone
	Maurice Limestone
	Gros Ventre Fm.
0'	Gallatin Limestone

PRECAMBRIAN

Age scale:
- 0 — Quaternary (CENOZOIC)
- 50 — Neogene / Paleogene
- 100 — Cretaceous (MESOZOIC)
- 150 — Jurassic
- 200 —
- 250 — Triassic
- Permian
- 300 — Carboniferous (PALEOZOIC)
- 350 — Mississippian
- Devonian
- 400 —
- Silurian
- 450 — Ordovician
- 500 — Cambrian

PRECAMBRIAN

Herbaceous Cretaceous

EPISODE	AGE:	FORMATION:	ANCIENT ENVIRONMENT:
8	**72 MILLION YEARS, CRETACEOUS**	**MEETEETSE FORMATION**	**WARM AND WET**

Past

A storm brews over a broad meadow as a small herd of hadrosaur dinosaurs grazes peacefully. These animals are calm and happy – they have plenty to eat and a wide-open vista so they can watch for approaching predators. The meadow is lush and varied. Palmettos grow among thickets of dense ferns in the broad, open patches, while scrambling broadleaf vines crowd the banks of the small streams that cross the pastoral landscape. In muddy low spots, a diversity of small cycads and fan-like ferns form a distinctively different patch. Conifer thickets appear in the distance.

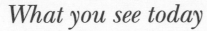

What you see today

Like the rest of the formations in the Bighorn Basin, the Meeteetse preserves a series of buried ancient landscapes – but it does it in a special way. Between 76 and 68 million years ago, volcanoes were regularly erupting to the west, and volcanic ash rained down on the landscape, burying entire ecosystems. So while it is not typical to preserve ancient herbaceous vegetation, it happened here. In 1990, Scott Wing from the Smithsonian Institution discovered the Big Cedar Ridge fossil site while prospecting for fossil plants.

He quickly realized that he had discovered the Cretaceous-plant equivalent to Pompeii. Over the next several years, he brought huge teams to the site, and they systematically excavated more than a hundred patches of the buried landscape. With so much data, Wing was able to identify three distinct types of herbaceous vegetation and literally map them onto the ancient landscape. This was the first time that anyone had ever truly reconstructed the vegetated landscape of dinosaurs.

Significance

There are only eight major groups of living plants: (1) flowering plants, (2) conifers, (3) cycads, (4) ferns, (5) horsetails, (6) *Gingko*, (7) an obscure group of forest-floor plants called lycopods, and (8) a very obscure group that includes the plant known as *Ephedra* (or Mormon tea). The flowering plants are by far the most common today and include everything from grasses to wildflowers to broadleaf trees to palms to water lilies. Flowering plants first evolved at the beginning of the Cretaceous Period, about 130 million years ago, and the Meeteetse site is one of the best places to document how and when flowering plants began to dominate the world's vegetation.

Far left: A backhoe excavated the volcanic ash horizon at the base of Big Cedar Ridge east of Worland, Wyoming.

Fossil leaves, from left to right: fern, lobed broadleaf, broadleaf, fern, palmetto, and cycad (sometimes called the "Wyoming Weird Plant")

Bottom right: Two teams from the Smithsonian excavate the base of the Big Cedar Ridge volcanic ash layer in search of in-place fossil plants.

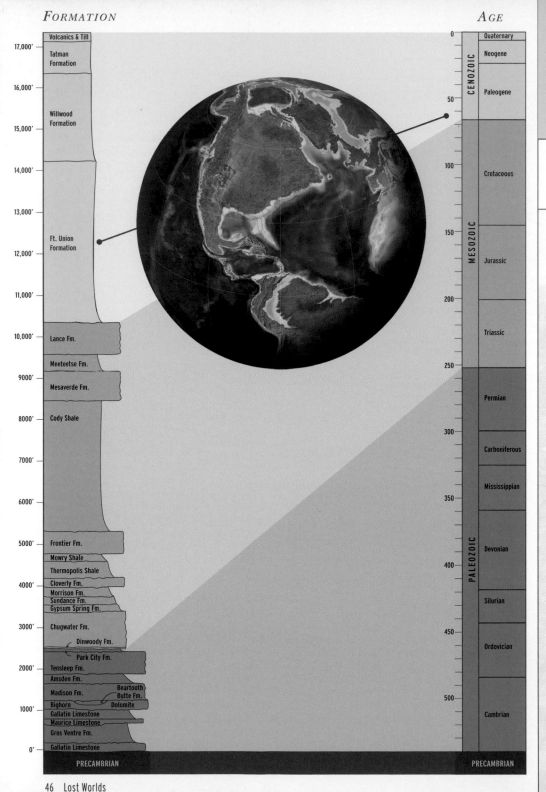

Mammal Swamp

EPISODE	AGE:	FORMATION:	ANCIENT ENVIRONMENT:
9	**57 MILLION YEARS, PALEOCENE**	**FORT UNION FORMATION**	**WARM AND WET**

Past

A dank and damp flooded forest grows right at the foot of the Beartooth Mountains. Reeds resembling cattails grow in standing water, and the water is full of crocodiles, turtles, and champsosaurs (alligator-like animals). Primate-like mammals known as *Plesiadapis* scramble around in the verdant foliage. The ground beneath the forest is spongy with the remains of dead trees. Nearby, roaring mountain streams rush down off the top of the recently formed Beartooth Mountain range.

What you see today

The area around Bear Creek, Montana, is littered with old mining equipment, and the remains of the Bear Creek Mine bear testimony to a horrible mine accident in 1943 that killed 74 miners. The mine never reopened and that was pretty much the end to coal mining in the northern Bighorn Basin. Less than 200 miles due east, and over the top of the Bighorn Mountains in the Powder River Basin, lie some of the world's largest coal mines. The coal in both basins formed around the same time, as warm climate conditions and uplifting mountain ranges created the perfect conditions for the formation of forested swamps. In the Bear Creek area, the coal seams were rarely thicker than 8 to 10 feet, but in the Powder River Basin some of the seams were more than 80 feet thick. The forests that were buried to form the coal were composed of conifer trees related to the living *Metasequoia* (dawn redwood) and *Taxodium* (bald cypress). Broadleaf trees were also present in these swamps. Conditions were warm and wet enough for the forests, but the formation of the Beartooth and Bighorn Mountains also played a role in the climate by creating enough local topography to alter rainfall patterns.

Right: Tilted layers of the Fort Union Formation are exposed in Foster Gulch near Lovell, Wyoming.

Significance

The coal in the Bighorn Basin was mined underground, while the big Powder River Basin coals are mined by stripping off the overlying rock. Coal accounts for 20 percent of Wyoming's economy, and the state produces 40 percent of all the coal burned in the United States. In the Paleocene swamps, carbon in the coal was concentrated by photosynthesis as the plants captured sunlight and used it to transform carbon dioxide into wood, bark, and leaves. Without these ancient, buried Wyoming forests, the modern economy of Wyoming would be much different. Burning the coal releases the carbon dioxide that has been trapped in it since the trees were alive nearly 60 million years ago. With coal, there is a direct connection between Wyoming's geology, the state's economy, and the changing composition of today's atmosphere.

Left: *Plesiadapis cookei*, an extinct tree-dwelling primate-like animal that is common in the Paleocene Fort Union Formation

Top right: A jawbone with teeth of *Ignacius clarkforkensis*, an extinct evolutionary cousin of modern primates that had a large set of front teeth

Bottom right: A petrified tree trunk exposed in the wall of a strip mine near Bear Creek, Montana, shows how a tree was buried by layers of silt in the late Paleocene.

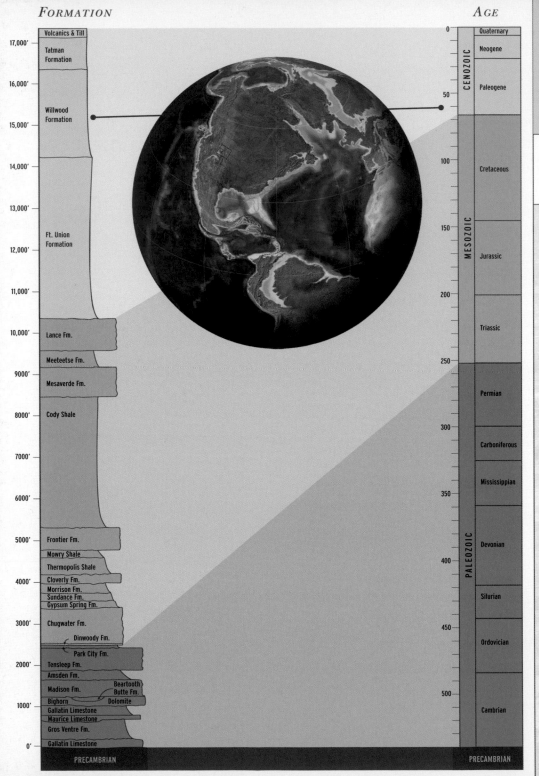

Greenhouse Bird

EPISODE 10	AGE: 54 MILLION YEARS, EOCENE	FORMATION: WILLWOOD FORMATION	ANCIENT ENVIRONMENT: HOT AND DRY

Past

The large, flightless bird *Diatryma* quietly moves through the lush floodplain forest, stalking a cocker spaniel-sized horse. The horse is startled by the rustling leaves and starts to run over to the fast-flowing stream cutting across the forest floor. Rainfall in the adjacent mountains feeds the streams here, and the temperature is much warmer than in today's Wyoming. This forest harbors an abundance of species – primates, tapirs, rodents, and crocodiles living amid laurels, legumes, and palms. It looks and feels like a subtropical ecosystem, yet Wyoming sits at just about the same latitude as it does today.

Below left: Paleontologist Ken Rose holds the startlingly robust lower jaw of a *Diatryma gigantea*.

Below: The badlands of the Willwood Formation east of Cody, Wyoming, are easily recognized by their red-and-whites stripes.

What you see today

The Eocene Willwood Formation is exposed as red-and-beige-striped badlands throughout the middle of the basin. It formed as the large Rocky Mountain ranges that encircle the basin – the Bighorns, Beartooths, Owl Creeks, and Pryors – continued to rise. With the rising and eroding mountains delivering a steady supply of sediment into the actively sinking basin, a great thickness of sediment accumulated during the Eocene epoch. Mud settled on the floodplains and sand filled the channels, burying the remains of the animals and plants that lived there. The Willwood Formation is one of the thickest geological units in the Bighorn Basin – as thick as 5,000 feet – and it preserves one of the most abundant and diverse suites of fossil land animals and plants known anywhere in the world. The most common mammals in the Bighorn Basin today (pronghorn antelope, horses, and even people!) can trace their ancestry back to fossils found in the Willwood.

Significance

The early Eocene, when the Willwood Formation was deposited, was a period of extreme global warming. Crocodiles lived above the Arctic Circle at this time, and the Willwood shows us that Wyoming hosted a whole array of animals and plants that are more typical of a tropical environment than the mid-latitude, continental interior that it actually was. How could the world get so warm so far away from the equator? Mainly because the concentration of atmospheric greenhouse gases such as carbon dioxide was much higher than it is now. It is also likely that large, powerful storm systems carried heat from the equator to the poles, bringing with them volatile weather conditions. Many scientists wonder if we are heading back to a greenhouse world like the Eocene as we continue to burn fossil fuels (like Fort Union coal) and release long-buried carbon back into the atmosphere.

Far right: A skeleton of the massive Eocene bird, *Diatryma gigantea*. It is unclear if this bird was a predator or an herbivore, but some recent studies suggest that it ate plants.

Above: A jaw from a *Hyracotherium*, an extinct, dog-sized horse

Background: Outcrops of the Willwood Formation in the McCullough Peaks area, east of Cody, Wyoming

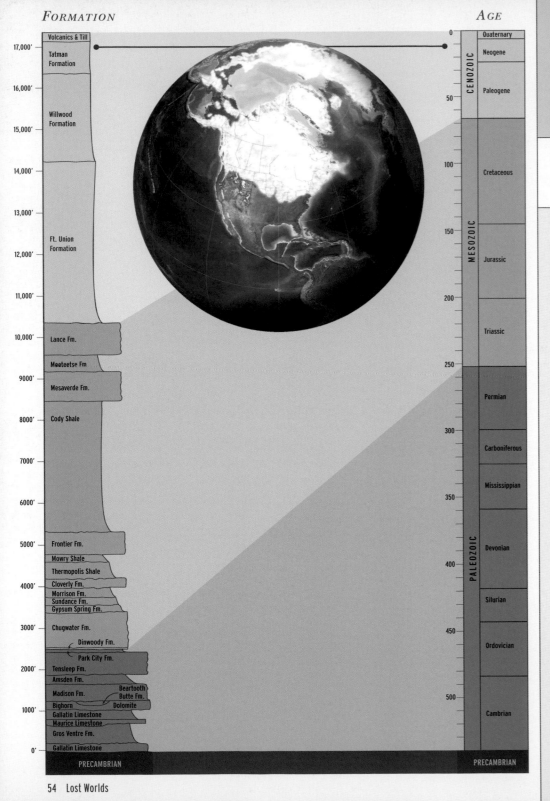

AGE

A Bad Day

EPISODE	AGE:	FORMATION:	ANCIENT ENVIRONMENT:
11	640,000 YEARS, PLEISTOCENE	YELLOWSTONE ASH	COLD AND DRY

Past

The three camels at the mouth of the Clarks Fork Canyon probably don't notice the strange cloud rising from the area of Yellowstone Lake some 100 miles to the west, although they certainly would have heard the loud explosion that preceded it. Within minutes the cloud will collapse under its own weight and roll eastward at speeds in excess of 100 miles an hour. The camels will have less than an hour to live. But the cloud will not stop there. It will continue to the east for several hundred miles, burning a path of death and destruction as it travels. Airborne ash will go farther still, covering most of the eastern half of the continent and smothering any life in its way.

What you see today

Today, Yellowstone National Park is one of the best-known natural sites in the world. More than 3 million people visit every year to vacation in its scenery, observe wildlife, and visit the geysers, boiling mud pots, and other thermal features that make this place so unusual – Yellowstone is home to more than half of the world's geysers. Geologists are growing more interested in Yellowstone as well. The thermal features suggest that there is great warmth in the ground beneath the park, and a series of seismic monitoring devices are now showing that hundreds to thousands of small earthquakes shake it each year. In 1959, a magnitude 7.5 earthquake on the western side of the park caused an 80-million-ton landslide that dammed Hebgen Lake and killed twenty-eight people who were camping along its shore. The devices that measure earthquakes form a network that lets geologists diagnose what is going on beneath Yellowstone, just like a surgeon uses a CT scan to look into a human body. Based on this data, it is clear that Yellowstone sits on top of a large cavity full of partially molten rock known as a magma chamber. The chamber begins about six miles below the surface and extends down at least 11 miles and is about 25

Left: A bison grazes peacefully while the Old Faithful geyser blasts superheated water into the September sky.

miles wide and 45 miles long. The last major eruption of this chamber occurred 639,000 years ago, and when it blew up, it released more than 250 cubic miles of molten gaseous magma and ash – more than 1,000 times larger than the 1980 Mount St. Helens eruption. The ash plume blew eastward and landed in measurable thickness as far east as Kansas City.

Significance

Evidence of ancient massive eruptions clearly has relevance for people who live within range of these volcanoes. An event that happened 639,000 years ago is not necessarily something we need to worry about, but it does make us think about geologic time.

Left: A skeleton of a large fossil camel from Nebraska. Bones of a similar species have been found in an archaeological site near Worland, Wyoming.

Below: The bright colors of the Grand Prismatic Spring in Yellowstone National Park are caused by microbes that thrive in high temperatures.

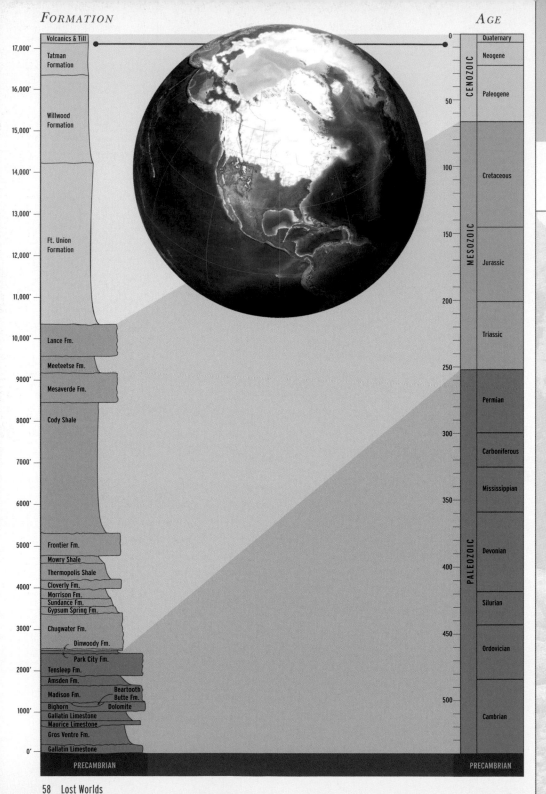

FORMATION		AGE

Volcanics & Till
17,000'
Tatman Formation
16,000'
Willwood Formation
15,000'
14,000'
13,000'
Ft. Union Formation
12,000'
11,000'
10,000' — Lance Fm.
9000' — Meeteetse Fm. / Mesaverde Fm.
8000' — Cody Shale
7000'
6000'
5000' — Frontier Fm.
Mowry Shale
Thermopolis Shale
4000' — Cloverly Fm. / Morrison Fm. / Sundance Fm. / Gypsum Spring Fm.
3000' — Chugwater Fm. / Dinwoody Fm.
Park City Fm.
2000' — Tensleep Fm. / Amsden Fm.
Madison Fm. / Beartooth Butte Fm.
1000' — Bighorn Dolomite / Gallatin Limestone / Maurice Limestone / Gros Ventre Fm.
0' — Gallatin Limestone
PRECAMBRIAN

Age scale: 0 Quaternary / Neogene / 50 Paleogene (CENOZOIC); 100 Cretaceous / 150 Jurassic / 200 Triassic / 250 (MESOZOIC); Permian / 300 Carboniferous / Mississippian / 350 Devonian / 400 Silurian / 450 Ordovician / 500 Cambrian (PALEOZOIC); PRECAMBRIAN

Dry and Icy

EPISODE	AGE:	FORMATION:	ANCIENT ENVIRONMENT:
12	**18,000 YEARS, PLEISTOCENE**	**PINEDALE MORAINE**	**FRIGID AND DRY**

Past

A thick river of ice emerges from the Beartooth Mountains and carves its way down Clarks Fork Canyon. The front edge of it is broken up and starting to melt. A snaking pile of rocks crosses the front of the glacier – debris bulldozed ahead of the ice as it moved through the valley. A short-faced bear strolls on the nearby hill, looking down hungrily at a mammoth and a few distant camels below. The bright-red stripe of the Chugwater Formation shows where the basin ends and the mountains begin.

What you see today

Clarks Fork Canyon has a river in it now, but it was sculpted in large part by glaciers during the mighty ice ages of the Pleistocene. During the last ice age, the entire Yellowstone Plateau was covered by a broad sheet of ice that fed these smaller valley glaciers around its margin. The boulder-covered hills at the mouth of the canyon are moraines, geological deposits that form from the rocks and debris that glaciers carry along with them as they flow along the landscape and then leave behind as they melt back. The moraines are like the gravestones of glaciers, marking their farthest position and leaving geologists with a rocky marker of their waxing and waning. Bones of the beasts that once roamed this area are sometimes found in these deposits. Mammoths, close cousins to modern elephants, and camels are no longer around, but grizzly bears still wander these parts looking for the next meal just as its short-faced cousin once did.

Left: A skeleton of the short-faced bear, *Arctodus simus*. This giant bear was much larger than a grizzly and had long limbs, suggesting that it might have been able run down large prey such as camels on open ground.

Right: A skull of the giant short-faced bear. Fossils of this species have been found in Wyoming.

Significance

Just as the greenhouse worlds of the Cretaceous and Eocene represent one climate extreme in Earth history, the ice ages of the Pleistocene represent the other. Coming and going on regular time intervals of a hundred thousand years, the ice ages seem to be controlled by wobbles in the Earth's orbit. How those wobbles precisely cause the buildup and melt back of so much ice on Earth is still being studied, but, as with the periods of extreme global warmth, the concentration of greenhouse gases also seems to play an important role. Many of the large mammals that inhabited this icy land-scape have since gone extinct, and some combination of postglacial warming and the arrival of new predators, humans in this case, was likely to blame.

Above left: This map shows the extent of the Yellowstone Ice Cap during the last ice age.

Above: A view looking west toward the mouth of Clarks Fork Canyon near Clark, Wyoming. The photo was taken from a glacier moraine, and the lightly forested ridge in the middle portion of the picture is a second moraine. The presence of the canyon and both moraines makes it quite easy to imagine this landscape when the glacier was present.

WHERE TO GO, WHAT TO DO

Visitor Centers and Museums in the Bighorn Basin

1. The Wyoming Dinosaur Center, Thermopolis. This museum runs an active nearby dinosaur quarry in the Morrison Formation where it is possible to pay to dig.

2. Washakie Museum, Worland. Recently renovated, this museum has an excellent overview of the geology of the Bighorn Basin, a full-size bronze mammoth, and exhibits of the paleontology and archaeology of the region.

3. Greybull Museum, Greybull. A small local museum with a long history and excellent samples of local fossils and minerals.

4. Bighorn Basin Research Institute, Greybull. A storefront in downtown Greybull with some local fossils and interpretations of local geology.

5. Draper Museum, Cody. A full-blown natural history museum that interprets the biology and geology of the Yellowstone Plateau and the Bighorn Basin and pays tribute to the basin's rich cultural history.

6. Bighorn Canyon Visitor Center, Lovell. This gateway to the Bighorn Canyon has an excellent movie and a three-dimensional model of the northern portion of the Bighorn Basin.

7. Cody Dam Visitor Center. Located at the western margin of the basin and at the edge of Rattlesnake Mountain, this visitor center provides great views of the Paleozoic part of the area.

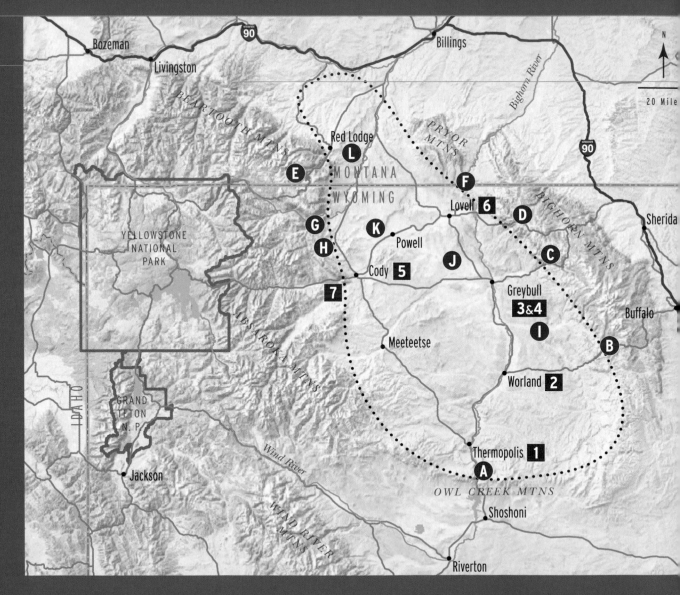

Road Segments with Key Viewpoints

A. Wind River Canyon (Hwy 789/20). The 40-mile drive from Shoshoni to Thermopolis to Worland is an amazing way to see the entire Bighorn Basin stratigraphy from your car window. The first 20 miles are in the Wind River Canyon with world-class views of the Precambrian and Paleozoic section. The second 20 miles are in the Bighorn Basin and show the section from the Chugwater Formation through the Willwood Formation near Worland.

B. Ten Sleep Canyon (Hwy 16). A beautiful road up a canyon with stunning exposures of the Paleozoic Formations.

C. Shell Canyon (Hwy 14). Another beautiful road up a canyon with spectacular exposures of the Paleozoic Formations.

D. Lovell to Burgess Jct. (Hwy 14A). Great views of the Bighorn Basin and vertically tilted Paleozoic rocks.

E. Beartooth Highway from Red Lodge to Cooke City (Hwy 212). A steep, winding road with amazing views from the top of the Beartooth Plateau and the best place to see the Beartooth Butte Formation.

F. Bighorn Canyon National Recreation Area (Hwy 37). Superb views of the Bighorn River as it carves into a deep canyon of Madison Limestone.

G. Clarks Fork Canyon near Clark. This is the best place in the Bighorn Basin to see the whole story. In one sweep of the eye, you can see from the Precambrian to the Pleistocene.

H. Dead Indian Hill (Hwy 296). A great place to see grand vistas and wonderful exposures of the Chugwater Formation.

I. Red Gulch/Alkali Byway (Shell to Ten Sleep). This gravel road passes by the Red Gulch Dinosaur Tracksite where you can see very clear dinosaur tracks in the Sundance Formation. Do not travel when wet.

J. Whistle Creek Road connects Hwy. 16 and Hwy. 20, west of Emblem. This gravel/dirt road is a splendid way to see the vast badlands of the Willwood Formation from a series of magnificent overlooks. As a bonus, there are lots of wild horses. Do not travel when wet.

K. Polecat Bench (north of Powell on Hwy. 295). This flat-topped surface is carved into the slightly tilted exposures of the Fort Union Formation and the Willwood Formation. Great view in all directions.

L. Belfry to Red Lodge, Montana, on Hwy. 308. This road winds through the Fort Union Formation and passes through historic Bear Creek, a coal-mining town. A plaque west of town commemorates the tragic Smith Mine disaster of 1943.

Museums Elsewhere with Good Wyoming Fossils on Display

Denver Museum of Nature and Science, Denver, Colorado

Museum of the Rockies, Bozeman, Montana

Casper College Tate Geological Museum, Casper, Wyoming

University of Wyoming Geological Museum, Laramie, Wyoming

The Field Museum of Natural History, Chicago, Illinois

American Museum of Natural History, New York, New York

University of Michigan Museum of Natural History, Ann Arbor, Michigan

National Museum of Natural History, Washington, DC

Yale Peabody Museum, New Haven, Connecticut

Natural History Museum, Western Wyoming Community College, Rock Springs, Wyoming

Suggested Reading

Carson, Robert J. 2010. East of Yellowstone: Geology of Clarks Fork Valley and the Nearby Beartooth and Absaroka Mountains. Sandpoint, ID: Keokee Books.

Fritz, William J., and Robert C. Thomas. 2011. Roadside Geology of Yellowstone Country, 2nd ed. Missoula, MT: Mountain Press Publishing Company.

Johnson, K. R., and R. Raynolds. 2006. Ancient Denvers: Scenes from the Past 300 Million Years of the Colorado Front Range. Golden, CO: Fulcrum Publishing.

Johnson, K. R., and R. Troll. 2007. Cruisin' the Fossil Freeway: An Epoch Tale of a Scientist and an Artist on the Ultimate 5,000-Mile Paleo Road Trip. Golden, CO: Fulcrum Publishing.

Love, J. David, John C. Reed, Jr., and Kenneth L. Pierce. 2003. Creation of the Teton Landscape: A Geological Chronicle of Jackson Hole and the Teton Range. Moose, WY: Grand Teton Natural History Association.

Authors

Kirk Johnson (right) is the Sant Director of the Smithsonian National Museum of Natural History. He is a geologist and paleontologist, and first discovered the wonders of the Bighorn Basin when he was a student at the Yellowstone Bighorn Research Association field station in Red Lodge, Montana, in 1981.

Will Clyde (left) is a professor of geology at the University of New Hampshire. He is fundamentally interested in Earth history, and has been studying the geology and paleontology of the Bighorn Basin since 1991. He recently led the Bighorn Basin Coring Project, a scientific drilling project aimed at building a better understanding of past periods of global warming.

Acknowledgments

We dedicate this book to the memory of the late H. Richard "Rich" Lane. Rich was a Sedimentary Geology and Paleobiology program officer at the National Science Foundation, and he believed in the value of science communication. In this role, he awarded the grants that supported the creation of the paintings in this book, and those in the book *Ancient Denvers: Scenes from the Past 300 Million Years of the Colorado Front Range*. We also thank the federal and state land managers and private landowners for their support of scientific research in the Bighorn Basin.